CANALS IN BRITAIN

Tony Conder

Published in Great Britain in 2017 by Bloomsbury Shire (part of Bloomsbury Publishing Plc), PO Box 883, Oxford, OX1 9PL, UK.

1385 Broadway, 5th Floor New York, NY 10018, USA.

E-mail: shireeditorial@ospreypublishing.com
www.shirebooks.co.uk

SHIRE is a trademark of Osprey Publishing, a division of Bloomsbury Publishing Plc.

A CIP catalogue record for this book is available from the British Library.

Shire Library no. 830. ISBN-13: 978 1 78442 050 5

PDF e-book ISBN: 978 1 78442 108 3

ePub ISBN: 978 1 78442 107 6

Tony Conder has asserted his right under the Copyright, Designs and Patents Act, 1988, to be identified as the author of this book.

Typeset in Garamond Pro and Gill Sans.

Printed in China through World Print Ltd.

17 18 19 20 21   10 9 8 7 6 5 4 3 2 1

COVER IMAGE
Front cover: Narrow boats on a canal somewhere in Britain (iStock). Back cover: The plaque from Hodgsons Bridge No. 157 on the Northern Reach of the Lancaster to Kendal canal (Alamy).

TITLE PAGE IMAGE
Lock 33 and Bridge 54 on Huddersfield Narrow Canal (Flickr Commons licence, Tim Green).

CONTENTS PAGE IMAGE
Coal at Cromford wharf arriving in bulk by water ready for horse-and-cart transfer to industry and the domestic market.

ACKNOWLEDGEMENTS
Alamy, page 28; Bridgeman Art Library. pages 16–17, 17 (upper); Canal and River Trust, pages 2, 4 (both), 5 (lower), 9 (upper), 15, 19, 23, 25, 26 (lower), 30 (lower), 31 (lower), 35, 37 (upper), 38 (lower), 39 (lower), 40 (lower), 41 (upper), 42 (upper), 43 (lower), 44–45, 46 (upper), 48, 51 (both), 53, 55, 57, 58, 59 (both), 60, 61, 62, 65, 66, 67, 68, 70 (both), 74, 75; David McDougall, pages 5 (upper), 8 (lower), 13, 20, 21, 22, 24, 26 (upper), 31 (upper), 32 (both), 33 (lower), 34 (both), 38 (upper), 39 (upper), 40 (upper), 41 (lower), 43 (upper), 46 (lower), 52 ,56 (both), 64, 77; Hugh Conway Jones, page 14; Mike Clark, pages 8 (upper), 9 (lower), 27 (lower), 47, 64 (lower), 71, 72, 73; Shropshire Union Flyboat Preservation Society, page 50; Tony Hurst, page 27 (upper); all the rest are from the author's collection. Also, thanks to Simon Salem and Marilyn McDougall for their assistance.

Shire Publications is supporting the Woodland Trust, the UK's leading woodland conservation charity, by funding the dedication of trees.

CONTENTS

INTRODUCTION   5

EARLY HISTORY   11

THE RIVERS ARE IMPROVED   15

BUILDING THE CANALS   20

CANAL MANIA   29

WORKING THE CANALS   36

THE RAILWAYS ARRIVE   49

VICTORIAN OPPORTUNITIES   55

WAR AND THE DEPRESSION   63

NATIONALISATION AND BEYOND   69

FURTHER READING   77

PLACES TO VISIT   78

INDEX   80

INTRODUCTION

TODAY, BRITAIN'S WATERWAYS are generally a tranquil, green environment largely cut off from industry. With clean water they are often havens for wildlife, and their towpaths offer a green ribbon into the heart of cities and towns. Some areas have been redeveloped, with housing around docks and wharves, but much of the landscape remains as it has been for years. Museums and interpretation centres on the waterways help us to make sense of centuries of waterway history, and some of the original buildings and historic boats have been preserved.

The first canals were raw scars across the landscape 250 years ago, dug to link coal mines and quarries to factories, and seaports to inland towns. They helped turn Britain from a largely agricultural nation into an industrial powerhouse ahead of the rest of the world.

The canal age, when waterways were virtually unchallenged as an inland transport network, was relatively short, lasting perhaps eighty years; in many cases less. However, most canals have a history of over 200 years and navigable rivers were first improved over 400 years ago.

What may look on a map like a designed network of waterways was actually a series of smaller initiatives built to join local industry to local raw materials. Many Acts of Parliament were created for canals between 10 and 15 miles long. The Leeds to Liverpool Canal was an exception: at 125 miles, it was the longest built.

Waterway development started with improvements to rivers and moved on to canal building. The first major project of the waterway age was the Aire and Calder Navigation, the first part of which opened in 1701, linking Leeds to the River Ouse and the sea at Hull. Built by merchants for merchants,

Opposite:
The National Waterways Museum, Ellesmere Port, incorporates a range of waterways from narrow to the biggest and has an excellent boat collection.

The Canal Museum, Stoke Bruerne, where you can explore the locks and walk to Blisworth Tunnel.

its success would promote the benefits of waterway transport for generations to come.

In the Midlands the first canals were independent waterways, begun between 1766 and 1769 but opening at different dates. They linked to one another and via the rivers to the great estuaries, so that through them the Midlands could connect to the sea. These early canals would make a fortune for their investors and owners.

In the north the success of the Aire and Calder was followed by the promotion of the canal from Leeds to Liverpool, which would link all the industries and resources of the region. A network of broad waterways stretched into coalfields and to the growing manufacturing towns.

The Grand Junction Canal was planned through the village of Stoke Brewen, as it was then known, in 1793.

Many of the Midlands canals were built to small dimensions. The boats on them could carry much less than the barges trading around

Stoke Bruerne
in horse boating
days in the 1920s:
a pair of boats
head south.

the country on the river navigations. There were rival broad canal schemes suggested in opposition to the narrow canals. On the Trent and Mersey canal, for instance, a wide canal linking the rivers Trent, Severn and Weaver failed to gain the political and financial support required, and the narrow canal was built. The smaller canals were seen to be affordable and a practical solution, while the general benefits of canals were still being evaluated. There was no overall plan to produce a national transport system.

The size of the narrow canals would be one factor in the long-term failure of waterways. It meant that only a maximum

The Gloucester
Waterways
Museum at the
heart of the docks
tells the story of
the Severn and
its waterways.

Leeds quickly became a huge city, thanks to the industrial growth supported by the Aire and Calder Navigation.

Even on a broad canal like the Grand Junction the narrowboat was important; two boats could share a horse and a lock, improving the economics.

of around 27 tons could be carried by a crew of two or three people. This was fine for the early days of industrialisation, but no competition for railway freight trains or trucks when they came on the scene. Despite years of discussion most of Britain's waterways never moved to a barge standard where hundreds of tons could be carried economically.

There was also no standardisation; lock lengths and widths even varied between narrow waterways. Where the narrow canals met the broad canals, 70-foot-long narrowboats could not pass through 60-foot-long locks designed for shorter and broader sailing barges. Broad waterways themselves varied from area to area, usually based on the local coastal barge size.

Another major factor that affected canals was the ability of companies to get on with one another and understand that they should get together to quote through-rates for goods moving long distances. The

'small company' mentality, linked to short-distance traffic, meant canals found it hard to compete on cost and flexibility with railways when they developed.

Britain's waterways were neither the first developed nor the largest in the world. The art of navigation was well known in China and Europe. Major artificial waterways opened elsewhere long before the British canal age. Unlike most of their European counterparts most of Britain's waterways were not funded by the monarchy or the state. They were built from private capital in a country that generally let the market decide what was best. European rivers were long and wide, giving long-distance trade routes. British rivers were short and fast-flowing, and development was already constrained in valleys full of factories and towns where land values were high.

At the point where canal owners might have invested to widen and develop their waterways in the 1830s the railways were built. As a result Britain's waterways would largely remain trapped in the size decided in the 1760s. It is perhaps surprising, therefore, that trade lasted so long and that we still have the canals today.

Barges and narrowboats meet at Preston Brook in the 1930s, where the narrow Trent and Mersey Canal met the wide Bridgewater. Cargoes would be trans-shipped between them.

Navigation improvements in Germany at Bahrenburg in 1724.

EARLY HISTORY

WHEN THE ROMANS occupied Britain in AD 44 they became the first rulers of the country to organise an economy and look at England as a whole. Roman rule saw salt mined in Cheshire, coal dug on the Tyne and wheat grown in bulk in East Anglia. Where there had been a local industry the Romans expanded it. Their trade routes used rivers and the sea and set the pattern of transport for bulk cargoes over the next 1,200 years. Many Roman towns were built at the highest point that tides ran to on a river, giving a navigable depth to the sea.

In East Anglia they created a network of rivers and artificial waterways. Development of the Fenland for agriculture saw the need for improved drainage. Canals were dug between Lincoln and Cambridge, the Car Dyke and Foss Dyke protected against flooding, intercepting the rivers and streams flowing into the fens and carrying boats. Branches connected to them included the navigable lodes and drains south of the Cam around Reach, which have navigable histories of over 1,600 years; some of the Roman waterways are in use today. Near Cambridge, Reach Fair received a royal charter in 1201 and the three-day trading extravaganza helped supply the city's needs. Boats used Reach as a port until around 1800; Reach Lode is still open for pleasure traffic.

Rivers came into their own for heavy cargoes. Quarries such as those at Barnack in Cambridgeshire, dug first by the Romans, continued to be worked for the best stone until nearly 1500. The stone was used for Ely Cathedral, Bury St Edmunds Abbey and other monasteries, and for buildings in Cambridge and Kings Lynn. It was taken by sledge to the River Welland and then moved by barges throughout the Fenland rivers and into Norfolk and Suffolk.

Ely Cathedral, above the River Ouse, was built of Barnack stone carried by water across the Fens.

A Roman waterway within the Wicken Fen nature reserve, linked to the River Cam, is still used for leisure today.

As an island Britain has always relied on its coastal shipping, and the great natural rivers – the Thames, Great Ouse, Trent and Severn – provided links far into the country. In the west in medieval times Bristol developed as the country's second city and the seaport for its area. Boats would come to Gloucester and Tewkesbury on the River Severn's tide; Worcester and Shrewsbury would also be supplied as river conditions allowed. Trade patterns included local and major fairs, the biggest of which would see traders buying and selling the raw materials of agriculture and finished products such as leather goods and woollens. Fairs would be timed to the likely availability of river transport in the spring and autumn, away from the floods of winter and the drought in summer.

After 1100, many rivers were gradually obstructed. From the records of disputes brought to the Kings' courts we can see the extent of navigation in Britain and its disruption. Medieval society demanded a lot from its rivers: drinking water, drainage, defence, power for mills, fish for food and transport by boat. An ever-increasing number of land-based activities caused problems to the navigations: weirs to hold back water for mill wheels blocked streams; fishing weirs with nets projected into channels; and fords for roads became more pronounced, producing new shallows. To aid navigation, mills would release 'flashes' or 'flushes' of water to help boats over the shallows. When water was short, however, mills were reluctant to release water and boats could be held up for days or weeks.

In 1249, Norwich, a major trading city, complained to the king that Great Yarmouth was blocking the navigation of the River Yare, preventing boats reaching the city as 'they used to do'. In 1270 the Corporation of Derby complained that the Derwent, 'navigable from ancient times', had been blocked by weirs built for mills.

The first Act of Parliament to improve a British river was for the River Lee in 1424 with a second Act granted in 1430. In Yorkshire an Act was passed for the River Ouse in 1462.

Boston on the River Witham: the outlet to the sea from Lincoln.

Witham, Boston.

A coastal sailing ship approaches Gloucester on the tide in the Middle Ages.

In 1553 John Trew of Glamorgan was engaged to build a canal to allow boats to pass obstructions on the local river up to Exeter. It was a small canal and carried boats of 16 tons shuttling between sea-going ships and Exeter's quays. Over the next 200 years it would be rebuilt and lengthened as a ship canal. On these early waterways individual businesses would have their own fleets of boats, independent barge masters picking up cargoes as they became available.

By 1600, the pace of life in Britain was changing. For the first time in 250 years the population had returned to the level before the Black Death and was reaching a point where improved agriculture was required to help feed it. Changes in the outside world would bring new prosperity and a new life to the economy. The spice trade route over land from Asia to Europe was cut off by the expansion of the Turkish Empire; following the lead of the seafaring Portuguese, the British too became explorers, finding new sources of raw materials and creating trade routes around the world. Inland navigation would link the interior of the country to the seaports and the world.

THE RIVERS ARE IMPROVED

IN THE YEARS before the English Civil War (1642–51) landowners began to look again to river transport as a means to develop their estates. A number of rivers needed improvement to get food into the growing city of London. To get over the weirs and fords blocking them money had to be invested, putting in locks and building bridges, so that natural waterways could become river navigations.

There were three ways in which a landowner could get a river improved. The ancient right was through the Commissioners of Sewers, bodies of local landowners working for the Crown to ensure that rivers were kept clear, primarily for flood prevention. They could help dredge a river but their powers did not allow for lock-building. Letters patent could be granted by the Crown to allow a body or an individual to work on a river, and after the Civil War parliamentary Acts became the normal route for permission.

Cropthorne water gate on the River Avon was removed in 1961 after over 300 years of use. The single gate controlled the depth in the river over many miles – a crude form of river improvement.

Not everyone was in favour of river improvement, however. Landowners with businesses away from the riverbank were worried about competition, and arguments frequently raged when a new scheme was proposed. Those against said that boats would break down bridges, watermen were known for stealing food and animals, and navigations destroyed fish weirs; those in favour argued that land carriage wrecked the roads and that rivers belonged to the Crown and had always been used. Even the Admiralty joined the arguments: rivers were important for transporting timber to the shipbuilding yards for the navy, and were a good nursery for sailors who might be brought into the navy at time of war.

As navigations were proposed deeper into the country, towns that had a monopoly on local fairs and markets fought against loss of business from imports. Reading, for instance, did not want to lose its corn market when the River Kennet improvement was put forward in 1715.

A gang of bow-hauliers works upriver with a barge on the busy River Thames. *The Thames at Twickenham* (oil on canvas) by Peter Tillemans (1684–1734).

Much of the work on rivers was done by gangs of men 'bow hauling' (pulling with ropes from the river bank) boats upstream. Sailing boats would use their sails as the wind allowed and then be hauled or poled forward. Some rivers like the Severn eventually passed Acts of Parliament and extended horse paths on their banks, putting the bow hauliers out of business. On rivers like the Stour in Suffolk horses were used but the towing path was so indistinct that a horse had 126

John Constable's *The Leaping Horse* (1825), one of his six paintings detailing the Stour Navigation. A rider urges a barge horse to jump a towpath obstruction.

hedges to jump between Sudbury and the sea, and on several occasions the horse had to jump onto the barge as the towpaths swapped from side to side with no bridge between them.

One area of concentrated activity was the valley of the River Severn and the rivers leading into it. William 'Waterworks' Sandys (1600–69), a landowner at Fladbury in Worcestershire, was involved with schemes to improve the

Avon and Teme. The Civil War interrupted his plans but after the restoration of the monarchy he was appointed to improve the rivers Wye and Lugg.

A major advocate for improvement, Andrew Yarranton (1616–84) was an active navigator and author. In his writing he described the cross of navigations necessary to join the major river estuaries of the country and provide inland navigation. An approximation of his scheme would be built by James Brindley around a hundred years later, forming the core of the English narrow canal system.

John Taylor (1578–1653) and Francis Mathew (active 1650–70) also wrote extensively about canal projects. London was largely supplied with coal by sea from Newcastle. Taylor and Mathew both proposed canal links or improved rivers to bring Forest of Dean coal to London and challenge the Northumbrian monopoly. Coal was one force behind river improvement as people began to turn to it as the fuel to warm their houses, cook their food and power their businesses. Limestone, which was pulverised to improve the land, and wool for textiles were other very important cargoes.

The first major strides forward in waterway development began in the north of England. In 1699 a bill was put forward to Parliament to make improvements on the River Aire from Leeds to Weeland on the tidal river. The bill was the result of the efforts of the wool manufacturers of Leeds and Wakefield and coal mine owners. Until the opening of the navigation they had had to rely on land transport to reach ships on the tidal part of the Aire or the Ouse. Changes in coal tax also made it a good time to put forward a scheme that would help Yorkshire coal reach the London market by sea. Larger ships could be used to cut costs, and the merchants would no longer have to pay so much for wagons and trans-shipment. They could cut their charges, sell more and still make bigger profits.

In the north-west, salt, coal and limestone were the key cargoes. In the 1720s a group of waterways was built around the Mersey and Ribble rivers. The River Douglas linked the Wigan area to the sea, helping the development

The River Medway improved between 1740 and 1800; the Admiralty threw its influence behind the navigation.

From the Middle Ages until the Canals and then the Railways provided a new transport system thousands of packhorses carried goods from coal to china cross-country on a daily basis. A boat pulled by a single horse could move up to 30 tons, whereas one packhorse could carry around 100 kg. In the background is Hawkesbury pumping engine on the Coventry Canal.

of the local coalfield and the export of coal to Liverpool and Dublin. The major engineer surveying and advising was Thomas Steers (1672–1750), who had built Liverpool Docks and is widely regarded as England's first civil engineer. Plans had been put forward to improve the River Weaver towards Winsford in the seventeenth century but nothing was done; but as coal replaced wood for heating the salt pans to dry out the salt, improved transport became important. At first, coal had come from Staffordshire, then Wigan, on the backs of packhorses – a laborious and costly business. When the new Weaver Navigation opened in 1732 it could deliver coal in far greater volume and take away the salt. As volumes increased and profits rose, investment in the area developed glass and chemical industries.

In 1755 an Act was passed to improve the Sankey Brook on the north side of the River Mersey. The waterway would link coal mines to the river network. It can be seen as either the last of the river navigations in the area or the first of the canals. The brook was so narrow that the company virtually built a parallel canal using the powers in their Act, thus getting round local opposition to artificial waterways.

Navigations did not have it all their own way; road competition with the developing network of improved rivers helped keep trade charges under control and ensured that the gains made by industry in cheaper fuel and transport costs were not lost to a monopoly.

BUILDING THE CANALS

By 1750 THERE had been major improvements on most of the major English rivers, owing to a whole range of contributing factors: increased wealth from industry; improved agricultural methods and a run of good harvests; a settled society; and increasing knowledge in science and medicine. The English Civil War had broken the absolute power of the monarchy and this had helped widen decision-making and allowed more individual freedom. Religious tolerance beyond the Church of England allowed non-conformists to become part of society and from that community emerged many of the new business class.

The Duke of Bridgewater was the sole investor in the Bridgewater Canal. He and his trustees built a network of canals and docks, which influenced and controlled canals in the north-west long into the railway age.

Inventions could be exploited by people with money to invest. Learned societies discussed how to improve society and used the relative freedom of expression and religious tolerance to put forward social theories and ideas to advance the country. As a result of better food supply more babies survived and the population grew. New factories drew people to the towns and this in turn required improved transport to feed the industrial areas. Britain found itself free to develop while much of Europe remained bound by conservative monarchies and absolute rule. People looked at the improved rivers and began to suggest schemes of artificial navigation, including canals to increase the reach of water transport.

James Brindley, the pioneer canal engineer.

There had been ideas for canals before but now there was money available to build them and there was a real demand for transport. Canals were not the only transport improvements being looked at: turnpikes (toll roads) and roads for horse-drawn trams were also being built. Some complemented the canals and rivers, bringing goods to their banks, while others competed. Round the coasts a vast merchant fleet sailed from port to port.

The first canal of the new age was the Bridgewater Canal. The 3rd Duke of Bridgewater succeeded to the title and the family estates near Manchester in 1748 when he was twelve. It was not until 1757 that he was allowed to control his fortune; two years later the Act of Parliament for his canal was passed and work began.

The Duke had taken the Grand Tour of Europe and then lived in London. Unable to marry the woman of his choice he retired to his estate at Worsley in 1758. On the Tour he had asked to see the impressive Canal du Languedoc – a wide waterway across southern France that had opened in 1681. The estate at Worsley had vast coal reserves. Previous schemes to link them to the river system had not been taken forward but now the Duke would build a wide waterway linking his coal mines to Manchester, similar to the waterways he had seen in France.

The final plan saw it sweep south from the mine, crossing the earlier Mersey and Irwell Navigation and running into the heart of Manchester. The aqueduct over the River Mersey,

The entrance to the underground canals into the coal mine at Worsley on the Bridgewater Canal.

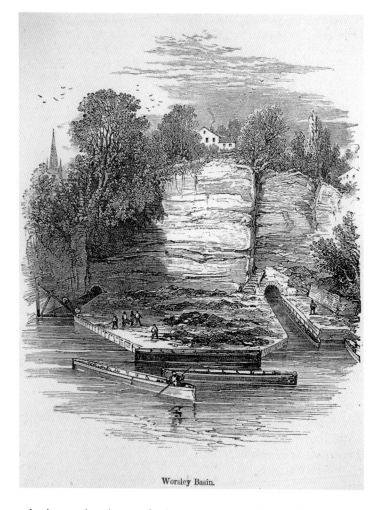

Worsley Basin.

which gave headroom for barges to sail underneath, became a tourist attraction for those with time and money: 'perhaps the greatest curiosity in the world … crowds of people, including those of the first fashion, resort to it daily'. Construction was overseen by the Duke, his agent John Gilbert and a millwright and engineer, James Brindley (1716–72). Described as 'the most extraordinary thing in the kingdom', the canal was a thoroughly modern undertaking: wide with no locks, it smoothed the contours of the land with embankments and cuttings. At Worsley the canal went directly to the coalface; once complete there were 48 miles of underground canal

on three levels within the mines. In Manchester the first warehouse was built over the canal and hydraulic hoists raised goods to the streets above. The boats on the waterway used an early form of containerisation. As soon as the canal opened passenger services also began.

More important than any of these developments, the price of coal was reduced by half – a message that would ring round a country where steam engines were being developed to perform more and more functions and coal was quickly becoming the most important fuel to drive them.

James Brindley would be at the heart of the next steps in the development of canals. South of Manchester and between the Mersey and Trent lay the Staffordshire potteries and coalfields; as early as 1758 plans had been put forward for a waterway to link the area to the navigable River Trent. Josiah Wedgwood, a leading potter, became the main voice in a campaign to provide the area with better transport links, in terms of both roads and waterways.

Initially known as the Grand Trunk, the Trent and Mersey Canal was promoted by Wedgwood's war cry, 'Your country calls'. He needed a coalition of people with money to raise the sum for a huge investment in water transport. His vision was for a waterway trust to build and run the canal, not for profit but for the public good, using the income to support the

Canals allowed the pottery industry to expand, bringing in raw materials when local supplies were exhausted. This is the Burslem Arm of the Trent and Mersey Canal in the 1930s, which was closed by subsidence in 1961.

Josiah Wedgwood.

Josiah Wedgwood worked hard to get transport links built to his pottery.

canal in the same way that turnpike trusts ran the new road system. If this had happened, then canals – like the roads – could have come into public ownership in the 1880s and their history would have been very different. As it was, Wedgwood and his partners set up a business raising the money from shareholders rather than risking a trust which might have fallen into the hands of a few rich men. The pattern of funding new waterways was set. The Act for the canal was passed in 1766 and the waterway opened in 1777.

There had been others looking at the same route: one serious proposal had been for a barge-width waterway from the River Weaver to the Trent. The greater influence of Wedgwood and his political allies won them the right to build. James Brindley took the unprecedented decision to build a narrow canal with locks 7.3 feet wide and 73 feet long, apparently based on the boats using the Bridgewater Canal from the mines at Worsley. The new canal would be cheaper to build than a wide one, and no doubt seemed quite adequate in increasing the volume of goods which could be carried. Eventually this decision would straitjacket the canals in their future competition with railways.

The promotion and building of the Trent and Mersey Canal spurred on a host of new waterway schemes. A link to the River Severn at Stourport from Great Haywood became the Staffordshire and Worcestershire Canal, opening up the south-west. The Coventry Canal was proposed to bring Staffordshire coal to Coventry and take products to the Trent and down to the sea for export. The Oxford Canal was then put forward to provide a link from the Coventry Canal going south to the Thames and the port of London. Birmingham had hoped for a branch from the Trent and Mersey but its first canal passed through the Black Country to Wolverhampton and added a link down to the Staffordshire and Worcestershire Canal. A further route north to the Coventry Canal opened in 1789. With water

The junction of the Staffordshire and Worcestershire canal with the river Severn at Stourport in 1776. (p 566)

transport assured, the West Midlands could develop new, heavy industries. Before canals were built they had to concentrate on small specialist items like clocks and guns because of transport difficulties; now they had a new future.

Canal companies were usually formed from their shareholders. A treasurer would be selected and at least on the early canals he would hold the funds in his personal bank account. An engineer would be chosen and in turn they would engage site engineers and contractors to build lengths of the canal. Contracting was always a hazardous business, especially if the company couldn't get its funds in quickly from the shareholders. Labour for the canal would usually come from the local area. Over time specialist teams of bricklayers and tunnellers would travel from canal to canal. There were also influxes of workers from Wales and Ireland at times of agricultural depression when canal jobs offered new hope. The 'navigators' (or 'navvies') – men who built navigations – lived in temporary quarters on the line of a new canal. Often badly managed and badly paid, they had a reputation for roughness.

Stourport-on-Severn, a canal town created by the company where the Staffordshire and Worcestershire Canal met the River Severn.

A navvy laying setts with the tools of his trade scattered around him.

The Calder and Hebble Navigation at Wakefield. John Smeaton's river improvement created an inland port.

There were incidents of riot but these were probably no more prevalent than among workers in other industries at the time.

When the Oxford canal opened from Coventry to the Thames in 1790, narrow canals finally linked all four major estuaries. They were not the only canals under construction: broad waterways were also built. The Thames and Severn Canal opened in 1789, finally completing a link first proposed in 1610 and much discussed in the seventeenth century.

In 1770 the Calder and Hebble Navigation opened, linking with the existing Aire and Calder Navigation. It extended water transport to woollen mills up to Huddersfield and helped Wakefield retain its importance to the woollen industry. As steam power began to replace water power, coal carrying would become increasingly important. Also in 1770 came the Act for the Leeds to Liverpool Canal. It was conceived to carry the sailing keels of the north-east across the Pennines to Liverpool and the north-west. It was planned, rerouted and replanned many times with ideas for branches south towards Manchester and links to Preston and Lancaster. Some of the winding course represented the promoters' wish to tap directly into as many industries as they could. When it was finished it joined the two sides of the country from Liverpool to Hull, via its links to the Aire and Calder Navigation and the River Ouse at Goole. However, it was busiest at its two ends, supplying the Aire Valley in Yorkshire and linking the Lancashire coalfield

Still carrying coal after 300 years: a West-Country barge loaded with power station coal in the 1970s on the Calder and Hebble.

to the cotton towns and the industry of Liverpool. The canal suffered from the problems all waterways endured in this period: a series of wars fought overseas cost the country dearly, increasing national debt, causing inflation and redirecting labour to fighting, or into industry to support the army and navy. The American War of Independence badly affected money supply in Britain and work on the Leeds and Liverpool was halted for ten years. Ultimately the 126-mile canal took forty-six years to complete.

GARGRAVE.

The Leeds to Liverpool Canal, Britain's longest waterway and the key link in a chain between the east and west coasts. A horse-drawn cargo boat makes its way across the highest level of the waterway in the years before the First World War.

CANAL MANIA

THE EARLY CANALS were a success. They began to carry heavy tonnages and the industries on their banks prospered and multiplied. Shareholders in the Trent and Mersey and Staffordshire and Worcestershire Canals saw impressive returns from their investment as profits rose. It was not uncommon for shares to return dividends of 50 per cent year on year before the railways came.

As the country became richer more people wanted to invest, but buying into the old canals was virtually impossible as shares were sold infrequently and at a huge premium – perhaps up to four times their face value. The pressure grew for new schemes into which investors could put their money. By 1790, ideas for new canal routes were coming from all over the country. In 1793 around twenty new canal schemes had their Acts passed, but almost immediately the effect of the war with France was felt in the economy. By 1797 the boom was over, leaving over thirty schemes under construction in a very difficult economic climate, with rampant inflation and little available capital, especially when it was discovered that the initial cost of building had often been underestimated by the engineers.

Few of these waterways would prove to be good investments either for their shareholders or for industry. Some ran parallel to earlier canals, seeking to improve on them. One success was the Grand Junction Canal, designed by William Jessop, which was a wide alternative to the narrow Oxford Canal on the major route between London and Birmingham.

The Kennet and Avon Canal, engineered by John Rennie, was authorised in 1794 and opened in 1810 as a broad link from the Thames to Bath and Bristol. When the plans for the waterway developed it appeared to have good prospects

Caen Hill Locks on the Kennet and Avon Canal, restored to navigation and reopened by Queen Elizabeth II in 1990.

An eighteenth-century cartoon captures the spirit of canal mania as a committee plans 'the Andes Canal'.

and there were far more investors than shares available. The original group of investors had to set up a meeting to apportion shares and decide plans. To protect themselves from being overwhelmed by demand they advertised a meeting in Devizes to discuss a canal plan while they actually met in Bristol. The newspaper advertisement caused a rush of potential investors to Devizes, and after hours of crowds searching for a non-existent canal company meeting, the town clerk of Devizes called the crowds together to try to regain order. A rival

The opening of the Paddington Arm of the Grand Junction Canal in 1801, another part of the link from London to the Midlands bypassing the Thames.

canal was proposed which satisfied people's desire to have their name on a list but after a few years this scheme came to nothing.

Some of the early waterways carried on developing. Among these were the Birmingham Canal Navigations. From the original single line of canal its branches spread throughout the Black Country. Growing for over seventy years, the original 22 miles becoming 140 by the 1850s and carrying over 4 million tons of goods per year. The old main line of the canal was bypassed in the 1820s and further improvements continued in the 1830s.

William Jessop, C.E.

William Jessop (1745–1814), probably the busiest of all chief engineers working during the canal mania, designed and built mainly broad waterways.

In England the last of the major canals opened as the railway age began. When the Birmingham and Liverpool Junction Canal opened in 1835 there was already a competing railway from Merseyside to the Midlands under construction.

In South Wales canals developed to link the seaports to the coal and iron of the hills. Relatively short waterways

The long-discussed Llangollen Canal approaches the Pontcysyllte Aqueduct – a project planned by Jessop and completed by Telford.

A twentieth-century view of Bilston Steel works, served by the Birmingham Canal since the Victorian period and busy with day boats that meant the Black Country waterways had a special culture of their own.

SPRING VALE WORKS.

climbed higher than most English canals, with deeper locks. The landscape was difficult but water supply was not an issue, as many were fed by rivers at their top ends. They opened up new coal mines and ironworks and started a major export market for Welsh coal. Foundry and engineering projects could be supplied with raw materials and their products taken down to the ports. Canals from Swansea, Neath and Cardiff had all reached up their local valleys by 1798. They

Harecastle Tunnels on the Trent & Mersey Canal. The second tunnel (to the left), planned by Rennie and built by Telford, ran parallel to Brindley's original tunnel (right), which had been badly affected by use and subsidence.

Northern Entrance of Harecastle Tunnels.*

The Aberdulais Aqueduct in South Wales, where the Tennant Canal crosses the River Neath. The Tennant is one of the canals linking Swansea to inland coal.

were connected to a network of tramways linking them to more distant quarries and mines. Boats on the South Wales canals were of an individual type, shorter than narrowboats but wider. Many had no cabins, as they were all suited to short distance and intensive use.

In Scotland a variety of waterways were built. The most successful was the Monkland Canal. With an Act of Parliament being passed in 1770 it was completed in 1793 and served Glasgow for the next century as the major supplier of coal to the city. The Forth and Clyde Canal opened as a ship canal across Scotland in 1790; it was over twenty years in

Canal and Rectory, Gilwern

Gilwern on the Monmouthshire and Brecon Canal, around 1900. Many tramways connected the canal to quarries and mines.

The passenger boat *Linnet* on the Crinan Canal in Scotland, opened in 1801 but much repaired and improved over the next twenty years, as the initial work had been poor.

the building as money ran short. It became a major supplier to the industries of the central belt and was much used by fishing boats following the herring shoals from one coast of Scotland to the other. In 1822 the Edinburgh and Glasgow Union Canal opened from Edinburgh to Falkirk on the Forth and Clyde. The two canals offered an intensive passenger service between the cities until the railways began to open in 1840.

Thomas Telford, the engineer who dominated the last phase of canal building in Britain.

Thomas Telford F.R.S.

The Caledonian Canal was first surveyed in the 1770s; it was seen as a safe route for sailing ships through Scotland rather than the dangerous seas round the north coast, and was supported by the Admiralty as a route for warships. Delays and cost problems meant that it didn't open until 1822. The canal involved William Jessop at the end of his career and Thomas Telford (1757–1834) at the height of his. Telford was the engineer of choice for many of the late canal schemes; if not directly in charge, he often acted as a consultant. The Caledonian Canal was not a financial success: much of it had been financed by government, and it was nearly closed several times in its early history as politicians looked to save money on a white elephant.

In Ireland the busiest waterway was the Lagan Navigation, which predates most English canals. Developed between 1756 and

'Little Venice', where the Regents Canal from the London docks met the long-distance Grand Junction Canal, around 1820.

1794 it delivered coal from Tyrone to Belfast and became busy with materials for industry. As imported coal became more important the navigation became busier – remaining so until road traffic seriously challenged it in the 1930s.

Irish waterways, unlike those in mainland Britain, were mostly state sponsored. The Lagan's major competitor had been the Newry Canal. Built by the state, it opened in 1742 and was the first summit level canal to be built in Ireland or Great Britain. A summit level canal is one which crosses high ground rising to a summit then falling – previously waterways had always been supplied with water from higher ground. Summit levels meant improved water supply was required to keep them deep. Tyrone coal for export was again the driver. For success Newry also needed a ship canal, and this opened in 1769. Both waterways were extensively rebuilt in the early 1800s, by which time coal was being imported from the UK. Canals also linked Dublin to the Shannon and the south coast, helping to stimulate some industry but also carrying passengers, including many emigrants leaving the country.

Investment in waterways produced canals in England and in parts of Wales, Ireland and Scotland, which first met the needs of industry and then those of financial speculators. They were successful as they suited the requirements of their age. The earliest were very successful – building costs were in line with income and they gave good financial returns; the canals of the mania period scarcely paid their investors back, however, and the last canals made no money at all.

WORKING THE CANALS

ALL CANALS WERE designed to be worked by horses. They had towing paths, and ramps up to locks and over bridges, all of which were graded to make them easy for the horse. Walls were smoothed to allow the tow rope to slide over them without snagging. Where the towpath crossed the canal, 'turnover bridges' allowed the horse to pull onwards without the rope being detached. Some waterways had split bridges to allow the rope through a gap.

A horse at work on the River Lee in the 1950s.

To maintain a canal, companies established maintenance yards. Lock keepers and lengthsmen (men who looked after a 'length' of canal and its towpath) lived in cottages along the line of a canal and kept an eye on its condition. Lock gates and machinery would be prepared in the yard and the company's maintenance men would be gathered for a major rebuild of a bridge or lock. Lock keepers helped boats through locks and protected the water supply.

Horse-drawn boats on the Grand Junction Canal before the First World War. The boat horse wears a crochet cap to prevent the flies from annoying him and keeping him from work.

Most canals had a reservoir or river to supply their highest level. Without an adequate supply, boats could not move on the canal. Drought was always a threat in summer, and in winter canals could ice up. Any disruption to navigation was a problem, so maintenance staff would use ice-breaking boats

Worn bridge guards at Norbury Junction on the Shropshire Union Canal. Grit from the towpath picked up by the tow rope cut into the protective ironworks. Ropes might last two or three trips for long-distance boats.

CAPTAIN CLARKE'S BRIDGE, HYDE

A turnover bridge on the Macclesfield Canal allowing the horse to switch from one side of the canal to the other without detaching the rope.

until the thickness of the ice became too great. Once railways opened, merchants had an alternative route and might leave canal transport for ever; even so, the problems associated with ice were never resolved.

Along the canal, public houses offered refreshment to boatmen and stabling for horses. On passenger boats and fly boats (boats working non-stop), horses would be changed frequently. Spare horses would be kept at company stables or pubs. For slower boats, horses would be stabled overnight in canal-side stables and at pubs. Farriers' and blacksmiths' shops in villages along the canal were available to reshoe horses and carry out general repairs to boats and equipment.

Most early canal tunnels were built without towing paths. Horses would be led over a defined path to meet the boats at the other end of the tunnel. Meanwhile boats would be 'legged through' (men would lie on a plank and propel the boat forward using their feet) either by their crew or by professionals who waited by the tunnels for work.

The Gayton yard maintenance team at work in the 1970s at Stoke Bruerne locks on the Grand Union Canal. Bill Nightingale smoothes the lock gate with an adze as Stan Voyce looks on.

The canal maintenance yard at Norbury Junction on the Shropshire Union Canal (around 1910), a factory producing the artefacts necessary to keep a section of waterway open and the base for teams to work along 'the length' (the area designated to that maintenance yard).

Canal horses were generally quite small; many were second- or third-hand, having had harder work before they came to the waterways. In general they were well looked after, as without them the boatmen had no work. While working, most horses would have had a member of the crew walking with them to keep them to the job. Some horses knew the routes well enough to be left to work on their own ('baccering' in the boatmen's language). The steerer would have had a smacking whip to hand to crack in the air to keep the horse at work if it got distracted.

The winter of 1963 brought traffic to a halt from Christmas to Easter, trapping crews all over the canal system, just before economics brought carrying to a halt anyway.

Ice threatened industry once canal supplies were cut off. The boating families also suffered, as they could not earn.

Canal icebreakers manned by the maintenance yards worked hard to keep canals open and working in 1963. More than 10 cm of ice would defeat them, however, and canals would be closed.

A horse pulling 27 tons of goods and a 10-ton boat needed to keep a continuous pull on the rope to keep moving easily. The worst pull was starting off; for this, boatmen developed methods to help the horse, adding a pulley into the rope, for instance, to halve the immediate effort.

On the wider northern waterways there were horse-drawn barges and sailing boats. Sailing keels going inland to places like Sheffield would often leave their mast and sails at a depot and hire a horse and driver (a horse marine) to pull them on their journey.

Canal architecture is mostly Georgian on the early canals, with Victorian industry developing later. The first buildings, bridges and locks are often built in the local stone. Until a canal was open there was little long-distance transport of building materials other than around the coasts by sea.

The earliest waterways were built as 'contour canals', the easiest form of construction. They would follow a level on a hillside for as long as it was practical before rising or falling to the next level length. When the first line of canal was built in Birmingham, the

A boatman uses a pulley to turn the sharp corner at Fazeley as a day boat heads towards Birmingham off the Coventry Canal, around 1950.

engineer Samuel Simcock was accused of designing the canal using his initials to make it longer, and increase its profits. As engineering techniques improved, canals began to take on the landscape. The last major canal – the Birmingham and Liverpool Junction Canal, designed by Thomas Telford – runs from cutting to embankment, slicing through the landscape. Locks are grouped making them easier and cheaper to staff. These later canals were more expensive to build but quick construction was important as railways grew across the country, competing with the canals.

MONTON NEW BRIDGE.

A horse-drawn barge on the Bridgewater Canal, around 1910.

Standedge Tunnel and the nearby visitor centre, a good place to explore Britain's longest canal tunnel, perhaps three hours' solid legging for the boatman.

Between 1750 and 1830 canals were almost unchallenged. The new roads helped bring goods to them and ran to areas canals couldn't reach. As a waterway was completed its internal trade grew and longer-distance traffic would open up. Passenger services operated on most canals between population centres or brought goods and people into towns for market day. It was possible to pay to travel on a long-distance cargo boat right across the country, sharing the boatmen's cabin.

The Shropshire Union Canal; one of several high embankments carrying the canal across the countryside.

THE CANAL AT TYRLEY.   *Arnold, Photo., Market Drayton.*

The Shropshire Union Canal; a deep cutting leading to Tyrley Locks, around 1910. Lock flights were grouped for cheaper staffing.

Fellow's Morton and Clayton boat *Orange* sank in Blisworth Tunnel cutting in 1909. On the bank are the wagons bringing iron ore for the canal boats to carry to Hunsbury Hill Furnaces, Northampton.

Boatmen probably came from all walks of life: during the period that canals were built, Britain was involved in several wars, and as each peace was drawn up soldiers and sailors became unemployed and the canals were looking for workers. Agricultural employees who had lost their jobs through enclosure of land for sheep and crops joined them. Men who had helped build the canals would stay on to work the boats.

Previous
A passenger boat
heading for Paisley,
on the Glasgow,
Paisley and
Johnstone Canal,
shortly after it
opened in 1810.
The canal was
converted to a
railway in 1881.

Leggers lie out
'on the wings'
after bringing a
boat step by step
through Blisworth
Tunnel, around
1910. Legging
boards were
needed when
a narrowboat
came through a
broad tunnel.

Cartage firms like Pickford's moved onto the canals and offered regular services the length and breadth of the country. Boats would move foodstuffs, local letters, gold bullion and furniture from city to city. Other more local cargoes included manure and night soil (human sewage), bones to make glue and rubbish. Specialist craft were developed to carry animals from the farms to feed the cities; they would load hay for feed and corn to go the mills. There are records of regiments of soldiers

Canal Tunnel, Chirk.

Chirk Tunnel
was built with a
towpath: much
easier for the
horses once they
were used to
the dark.

being shipped by water. At Weedon in Northamptonshire a whole canal-side barracks was built to receive the contents of Woolwich Arsenal should the French invade.

The basic trade of canals was coal and limestone. From the earliest days to the last high-volume pans on the Aire and Calder Navigation in 2002, coal has been carried on every waterway in every type of craft.

Textile mills also came to the canal side. As business developed new engineering works supplied the machinery; coal, iron and steel were needed as well as the raw materials for the product – all these materials would be delivered by water.

Stone was very important too, both for buildings and for roads. As turnpike trusts developed across the nation, so stone from canal-side quarries was sent to build them. The Chesterfield Canal carried over 20,000 tons of building stone to rebuild the Houses of Parliament between 1840 and 1844 after the disastrous fire of 1834. Dartmoor stone carried on the Stover Canal to the River Teign in Devon was used to build London Bridge (dismantled in 1967 and reassembled in the Arizona desert) and the National Gallery. By 1830 there were 3,969 miles of waterways, around a third of which were narrow canals.

Three coal pans, each carrying 170 tons, are pushed through Stanley Ferry repair yard in the 1970s.

THE RAILWAYS ARRIVE

In 1816 the engineer John Rennie put forward a canal plan for a route including Stockton and Darlington in the coal-producing area of the Tees valley. There had been earlier canal proposals in 1767 which hadn't raised enough finance to start. When this plan finally developed into a line of transport it was not as a canal but a railway. When the Stockton to Darlington line opened in 1825 it was the world's first public railway with goods trains hauled by steam locomotives. Passenger traffic, at first horse drawn, became steam hauled in 1833.

When Richard Trevithick invented the steam locomotive in 1804 it had seemed just a toy, or that it might provide an enhancement to the tramways that served the canals and rivers. The early railways did not seem to offer any threat to the waterways, and the canals were in a strong position, as many had paid off their initial costs and could earn good profits. But, although the locomotives were not strong enough to pull heavy loads, they were cheaper to run than horses and used fewer staff. And at a time when the population needed to grow more crops, and when agricultural land was at a premium, they also saved on growing fodder.

In 1830 a railway opened between Liverpool and Manchester, with the Duke of Wellington, the Prime Minister in attendance. In 1833 Acts of Parliament were passed for railway lines from Manchester to Birmingham and from Birmingham to London. Railway development was swift and the technology advanced quickly too. Passenger traffic developed ahead of expectations, providing funding for development.

As more and more railways were planned across the country, canal companies began to worry. As local concerns they had not concentrated on the long-distance element of

The canal breaches near Warwick in the early days of the railway, with the new London to Birmingham line in the background.

The Shropshire Fly Boat Preservation Society's boat *Saturn*, which used to carry cheese speedily to Manchester, is preserved and attends rallies around the canal system.

their business and they were not necessarily used to working with their neighbours.

Initially railways were promoted in the same way that the canals had been, by local groups over relatively short sections. Through the 1840s these small companies amalgamated or were bought up to form much larger concerns like the Midland Railway. In 1846 Parliament acted to standardise the gauge of railways. The lack of standardisation of inland waterways and the small size of the narrow canals would prevent canals from competing efficiently with the railways.

In 1835, as railway competition was increasing, the Grand Junction Canal began to duplicate locks at Stoke Bruerne to speed up traffic. The pressure on locks could be tremendous with boats queuing to use them and fights breaking out over precedence. On one occasion at Stoke Bruerne a steamer towing a butty overtook the second of a pair of horse boats approaching Stoke Bruerne top lock. The leading boats of each pair entered the lock and neither captain would withdraw. Nothing moved through the lock for hours and the magistrate had to be called from nearby Towcester to resolve the problem.

Canal companies reacted to the railways in different ways. Many wanted to sell out even as the first competing routes were discussed, in the hope of getting the best price, in order

Between the two world wars, boats wait at Stoke Bruerne for the locks to reopen after maintenance, echoing the busy days of the canals when they would need to wait in turn.

to invest in the new railways. Not all railways had the capital to buy out the competition but the North Staffordshire Railway was happy to purchase the busy and profitable Trent and Mersey Canal when the offer was made in 1845: the railway would use the canal to supplement its traffic.

Some waterways became railway companies. For example, the Birmingham and Liverpool Junction Canal opened in 1835 only after receiving government loans to help complete its construction. By 1844 it was looking to amalgamate with the older and established Ellesmere and Chester Canal. In 1845 the

The North Staffordshire Railway bought the Trent and Mersey Canal in 1847, using it as an extension to their railway routes. The electric tunnel tug became operational in 1914.

joint company put forward the idea of turning the route from Wolverhampton north into a railway: building on the canal bed and making additional railway routes. The London and North Western Railway Company, building a route in parallel with the canal, moved to stop this competition and leased the waterways. This combine ran the waterways, which it controlled as the Shropshire Union Railways and Canal Company. Able to trade with its own boats on its own waters and backed by a railway guarantee, it developed into an efficient network offering timetabled boating services. Its boats could also penetrate the territory of rival railways and bring in new business.

The London and North Western Railway Company also tied up a deal with the important Birmingham Canal Navigations, leasing them from 1846 and guaranteeing their dividend. The canals of Birmingham and the Black Country would continue to develop but also serve the railway. Interchange basins opened around them, moving goods carried long distance by rail the last few miles on water and taking the finished product from factory to rail.

Most canals up to this point had merely provided the route for business; the trade itself was carried on by individuals or firms of carriers. As the railway age dawned and competition bit, Pickford's among others left the waterways for the railways. From 1845 a series of Acts of Parliament tried to help balance the struggle between canals and railways; one of these Acts allowed canal companies to become carriers themselves and several companies took advantage to maintain services for their customers.

Oldbury railway interchange basin, around 1910, where the Great Western Railway connected with the Birmingham Canal, linking long-distance rail to factories established on the canal bank.

1930s: six locks up the Runcorn flight. The locks lead the Bridgewater Canal down to a port by the Mersey and later the Manchester Ship Canal. Railways became the backdrop to waterways everywhere.

The Kennet and Avon Canal had long kept an eye on the railways, sending their engineer north in 1825 to see what was happening. There were already railway schemes in discussion between London and Bristol, and by 1833 the canal was proposing its own railway on the route. Even as the Great Western Railway's bill was discussed in Parliament in 1835 the canal reduced its tolls to minimise the loss of business and make the railway's financial predictions look wrong. The idea that steam carriages might run on the turnpike road from London to Bath and Bristol (making the railway less financially attractive to passengers) was also encouraging. However, the Great Western's bill was passed, the railway was built and opened, and canal trade was lost. A last idea of widening the canal towpath and building a railway on it but keeping the waterway open was also abandoned. In 1852 the canal was sold to the railway for just over £200,000, a fifth of the cost of construction forty years before.

Other waterways tried to gather support to resist railways. The Grand Junction Canal put together a committee of neighbouring waterways to establish better through working and to fight railway bills in Parliament. They failed to convince the important Oxford Canal to join in; the Oxford had never forgiven the Grand Junction for taking the major part of its London traffic. The Oxford had previously brushed off requests from the Grand Junction to widen the Oxford's northern route to Coventry – part of a vital coal supply route to London. In the 1830s, as railway building began, the Oxford had finally invested in this northern section, cutting out the old contour loops, shortening the line and speeding up trade but still remaining narrow. Railway competition would eventually cut into their profits but the Oxford was secure enough in its hold on trade and the diversity of its business to ignore the overall picture for canals.

Initially railways were seen as healthy competition for the canals by most people. Waterways had had the majority of inland transport to themselves for eighty years and were felt to be unhealthy monopolies, charging too much and needing a shake-up. It was thought that railways would compete and force canal rates lower, and that industry would benefit from cheaper transport. Between 1845 and 1847, 948 miles of canal – virtually a quarter of all waterways were bought by the railways. By 1850 there were 10,000 miles of railway across the country. Some canals had never reached their potential. Ironically their most profitable years would be spent carrying the materials which built the railways alongside them and which would effectively end their carrying life. Once railways were seen to be dominant and started to take over canals, people began to worry that railway rates would rise and that they would become a new monopoly. In 1855 the Canal Association was formed, led by the Aire and Calder Navigation. It promoted the development of waterways and watched railway legislation to look after canal interests.

Some canals were closed so that their routes could be used to build new railways. In the south the canal to Andover from the River Test near Southampton had been fought over by railways. The route was worth more than the waterway, which was not busy. To resolve matters the canal owners formed a railway company to take over and build the track themselves. The canal closed in 1859, the railway opened in 1865 and worked until 1967, but neither canal nor railway satisfied their shareholders' expectations.

The Aberdeenshire Canal in Scotland had always been marginal; it was so far north that it had to shut for the winter every year because of ice. Its course offered a good route for a new railway and in 1854 it closed as the Great North of Scotland Railway was built over it.

VICTORIAN OPPORTUNITIES

From the 1880s a series of government enquiries began to look at the state of the canals and what might be done to prop them up against the competition from the railways.

Independent waterways felt threatened by railway-owned canals. Between major cities there would be several independent canal companies. If one of these became railway owned the others worried whether the railway-owned canal might be closed, badly maintained or uncooperative when long distance freight rates needed to be quoted.

Railway competition stimulated the bigger waterway concerns into positive action. Up to 1840 the River Severn had been a purely natural river with no locks or weirs to help boaters. Some work had been carried out to replace bow hauling with a towpath for horses. Canal links had joined it over time – the Staffordshire and Worcestershire Canal in

A steam tug at the head of a train of empty compartment boats in the 1950s, shortly before diesel tugs replaced them. In front of the tug are the false bows which will go into the loaded train behind the tug to break the wash of the tug's propeller.

Lincomb Lock on the River Severn; navigation improvement was stimulated by railway competition.

1772 and the Worcester and Birmingham Canal in 1815 – but down to Gloucester they relied on the unimproved river, which nearly dried up in some summers over the shallow parts. The lower river had been paralleled by the Gloucester and Berkeley Ship Canal, and Gloucester had begun to develop as a port after 1827. From 1842 locks began to open on the river. By 1852 the Midland Railway and the Great Western Railway had both opened branches to the docks in Gloucester. Above Stourport the last barge passed to Shrewsbury in 1862, the year the rival railway opened. It was not until 1871 that the last lock opened at Llanthony to pass through the new weir built to improve the river channel through Gloucester. This lock allowed coal from the Forest of Dean up to Gloucester docks. Eight years later a railway bridge opened across the river at Sharpness and the main coal flows on the water from the Forest of Dean were finished.

In Yorkshire the Aire and Calder Navigation running through the coalfield was also determined to develop and

A loaded train of compartment boats pulls 700 tons of coal towards Goole and passes an empty train heading back up the River Don in the 1930s.

maintain its business. Its innovative engineer William Bartholomew widened and deepened the waterway – the third time this had been done since 1699. His son succeeded him and introduced railway technology to the canal, inventing the compartment boat system. Individual square boats could be pushed or pulled in trains, each boat carrying up to 40 tons. From the 1870s this settled down to around nineteen pans being pulled by one tug. The Aire and Calder Navigation maintained its links to collieries and offered to build new wharves when necessary. Pans were loaded directly at collieries and could be lifted and tipped into seagoing ships at Goole.

In the north-west the volume of traffic on the River Weaver and Trent and Mersey Canal was less affected by rail than elsewhere. Salt traffic on the Weaver was heavy; from 1871 locks were rebuilt to allow for even larger craft.

Steam also came to the canals. On the bigger waterways it worked well: boilers and engines reduced the tonnage which could be carried but the savings that could be made against feeding and running a horse plus the ability to work longer hours and tow unpowered barges more than compensated. Narrow canals also introduced steam but the loss of carrying capacity meant it was not widely applied.

Salt is loaded onto a steam barge on the River Weaver around 1900.

The smart all-male crew of the butty *Buckingham*, which would be towed by a steamer from London to Braunston, probably working to Birmingham behind a horse.

On all waterways the majority of carrying was by all-male crews. Many boats worked regular routes on contracts between mines and factories, and between factories and markets. There were also longer-distance boats on regular timetables carrying a whole range of goods. Alongside these were carrying firms who would go wherever the work required. Some boats were family operated and women were occasionally captains in their own right. As competition from railways cut into work and profits, more boats became family crewed; this was more common on the narrow canals. Usually the captain would be paid for the journey and it was up to them to pay their crew.

It was during this period that the decoration style known as 'Roses and Castles' came onto the narrowboats. All craft had had decoration; on the broad canals this was passed down from maritime or local traditions, and narrowboats featured company logos or slogans, but from the 1850s cabin panels began to be painted with scenes. Most boatyards had a boatbuilder trained as a sign-writer and they became the artists, as did some boatmen. The tradition stayed with the

A maintenance stoppage at Long Buckby in Northamptonshire allows a boating family to celebrate a christening. Major 'stoppages' were held at the bank holidays to cause as little disruption to trade as possible.

boats even when the newly nationalised fleet got corporate colours in the 1950s.

With long-distance firms like Fellows, Morton and Clayton developing nationwide after 1889, having families on board became common, whereas short-distance, regular trips and the longer-distance fly boats working day and night continued to be worked by men. Many families had houses ashore; others left their children with grandparents, returning home between trips.

In 1877 and 1884 legislation to improve living conditions was passed after a long campaign by George Smith, a social reformer. Cabins would be registered and inspected by local authorities to ensure they were watertight and not infested. Regulations also controlled who could share a canal boat cabin. On narrowboats forecabins at the front of the boat were added to take (usually) male teenagers displaced from the family cabin. Legislation was only partially successful, as it went against the boatmen's own wishes to pursue their

After a struggle, even the boats of the nationalised fleets were allowed what had by now become traditional decoration.

A narrowboat cabin looking from the double bed to the back of the boat; this is a family home.

business. They knew there was no compromise between the size of cabins and the amount of cargo: it was goods that paid the wages.

In 1888 the effect of competition with the railways, and examples of the possible modernisation of waterways such as the use of steam power and widening schemes were discussed in Parliament as it examined a Railway and Canal Traffic Act. In France the older canals were being standardised to take a minimum 300-ton barge. A plan was drawn up by the canal

Charles Nurser at work at Braunston, building and painting wooden boats.

interests for a 300-ton route to the heart of England via Hull and the River Trent. Further 100-ton barge routes linking to the Thames, Severn and Mersey were also planned, and this became the standard view of an improved system. It would have changed the canal system completely and many of the eighteenth-century canals we have today would have been built over, but there was never the parliamentary will or the money to make it happen, so it remained a dream.

In 1906 evidence began to be taken by a Royal Commission on Canals and Waterways, looking at their place in transport and at what could be done to improve them. Ultimately government decided it could not get involved because this would be unfair on the main railway companies.

The overall effect of railways on the waterways was to cut off new traffic and cut the rates they could charge, reducing the income from existing cargoes. All waterways had to cut their costs and their staff. For those that had never prospered, closure loomed. London lost its link to the south coast via the Wey and Arun Canal – an Act of abandonment was passed in 1868. Around the coasts many smaller isolated rivers and canals with limited traffic stopped carrying.

One of the ideas which came out of the Commission to take matters forward was to place the ownership of canals into the hands of a public trust for the good of the nation. But almost as soon as the Royal Commission had been completed, and had reported, the country was plunged into war in 1914, and all ideas of improvement faded into the background.

WAR AND THE DEPRESSION

Canals lost around 10,000,000 tons (equivalent to a third of their tonnage) during the First World War. The railway-owned canals were put under national control at the start of the war. This gave them some protection but the independent canals were left to operate as normal. There was still traffic offering but the east coast ports were largely closed to canal trade as they became naval bases and as shipping in the North Sea came under attack. Coal traffic on the Aire and Calder Navigation dropped by 90 per cent. The Forth and Clyde's eastern terminus at Grangemouth was closed.

Kitchener's plea for volunteers for the army saw boatmen and maintenance staff leaving the canals to join up. As industry geared up to fight the war the strain fell on transport. Around Britain new armaments factories opened, along with all kinds of businesses supporting the war effort. The pressures were such that canal transport was asked to relieve the railways but there were not enough staff left to take the strain. In 1917 the independent canals were finally brought under government control. Around 200 soldiers taken into the Royal Engineers were trained at Devizes on the Kennet and Avon Canal and they worked on waterways to help carry the goods. Over 1,000 soldiers were employed across all the canals. Nationalisation meant that canals had some financial support to make up for the loss of traffic.

All through the war, canal professionals worked to support the armies at the front. Some wide boats from the Grand Junction Canal were taken to France and worked alongside French barges on the canals carrying goods to the front. In the Middle East canal men carried goods on the Tigris and Euphrates rivers.

Wide boats at Brentford in the 1930s. They worked extensively up the Grand Union Canal towards Cowroast.

A busy Gas Street Basin in Birmingham before the First World War. This was where the Worcester and Birmingham Canal met the Birmingham Navigations.

At the end of the war the canals were in a bad state. Little maintenance had been done and many companies felt they needed to start all over again to bring business back. During the war factories built for the war effort had often been located alongside roads and railways. Men coming back from the war had new horizons and new training; people had learned to drive lorries and maintain diesel engines and there was a

Gas Street Basin in the 1960s after canal carrying had finished.

The 1920s saw industrial unrest and there were several boatmen's strikes as wages were cut. This is Braunston in 1924.

large number of war surplus vehicles that now carried goods in competition with canals and railways. However, the east coast ports reopened and the wider waterways of the north-east gradually reclaimed business.

As the canal bosses came back from the war, waterways were still nationalised. They pressed the government to continue nationalisation and take over the waterways permanently; the country needed to pay for the war, however, and could not afford to buy or pay to improve the canals. In 1920 a Committee of Parliament led by Neville Chamberlain recommended that canals should be returned to their owners.

As trade unions became more powerful they had a significant impact on the canals. Boatmen were used to using all hours of the day to work; boats were slow and the incentive of moving a cargo as quickly as possible to be able to load the next one was always there. Once restrictions on the working day came in after the war it had a profound effect on the canals, especially where the company was the main carrier (like the Shropshire Union Railways and Canals, for example) and where unions were strong.

After the war there was an initial industrial boom as home demand for ships to replace those lost in the war created work. However, the loss of European markets because of the disruption of the war and the increasing independence of the Dominions made life hard for exporters, and the economy stalled.

Railway and road competition was another factor, but in many cases waterways were also affected by their age. Factories built on them 150 years previously had closed or relocated to roadside or rail-linked sites. Old coal mines ran out of coal and quarries were worked out. Some canals built through coalfields were affected by subsidence as the land fell into old workings, bridge heights suffered and tunnels became impassable. Twenty-four canals would close or cease carrying in the period between the world wars, including main-line routes such as the cross-Pennine Rochdale Canal.

There were also improvements. The scheme to widen the River Trent (part of the 1880s plan for a new waterways network) went ahead. The Corporation of Nottingham pushed for the improvement of the river with new locks capable of taking a tug and three barges. Unfortunately the scheme was not taken on to the Black Country as originally envisaged and inland from Nottingham most cargoes transferred from 270-ton barges to 27-ton narrowboats for the last stage of the journey.

In 1920 the manager of the Grand Junction Canal felt that long-distance shipment of goods was likely to become a thing of the past. The failure to widen the waterways meant goods would travel by rail not canal. The Grand Junction's route became the Grand Union Canal in the 1930s with a group of

A new warehouse under construction for the Grand Union Canal at Sampson Road, Birmingham, as efforts are made to rebuild waterway trade between London and Birmingham.

waterways coming together to link London to Birmingham and Nottingham for the first time under one company.

As the government acted to quell social unrest and get people back to work in the 1930s, money went into the building of new wide locks from Braunston to Birmingham to develop a barge route for 60-ton boats across country. However, tunnels and the bed of the canal were not widened, which meant it was difficult for wide boats to pass. Instead of barges the company invested in 186 pairs of narrowboats, each pair consisting of a motor towing an unpowered partner, the butty.

They worked hard to get cargoes back from the railways. This bold plan worked in part but they could never find enough crews to put more than two thirds of the new boats to work, as too many of the links to boating families had been broken in the First World War. Canals and boating almost disappeared from public view; the old way of life was not compatible with the growing changes in society as freely available education brought about social change and new expectations for living conditions and work.

A pair of Grand Union Canal carrying craft move south from Blisworth Tunnel. The unpowered butty is pulled by a motor.

BRITISH WATERWAYS
HARECASTLE
TUNNEL

NATIONALISATION AND BEYOND

C ANALS WERE AGAIN brought under government control in 1942 as part of the war effort in the Second World War. Boatmen's jobs were given protection and it was difficult for them to join up. The government recognised that the canals could serve an important role in relieving the railways. One challenge was to get the goods away from the ports, as masses of material arrived on convoys; buffer depots were built alongside canals to take bulk supplies and hold them before issue to the shops.

Training schemes were introduced on major routes like the Grand Union Canal and the Leeds and Liverpool Canal and women trainees came to the canals to work. After the war their insights provided a lot of information on the boating families and the way canals worked.

In 1945 as the war finished the Labour Party was voted into power. Nationalisation of transport was one of its major policies and came into effect on 1 January 1948. Some canals were not taken over, such as the Rochdale Canal, which survived as a property company and car park, not a transport business. The highly successful Manchester Ship Canal also remained independent. In the early days of nationalisation some canals remained with their railway owners and the canals still ran at least one railway. Some canals reacted violently: the Staffordshire and Worcestershire did not want to be nationalised and it is rumoured that the company put most of its paperwork on a protest bonfire.

Change came slowly; national finances were thin as the war had to be paid for, and the government had taken on a huge programme. In 1953 the British Transport Commission began an enquiry under Lord Rusholme to make recommendations on managing their waterway assets, and his

Narrowboats of the British Waterways northern fleet enter Harecastle Tunnel in the 1950s from the south through the ventilation system installed in the diesel-powered age.

Wigan Power Station on the Leeds and Liverpool Canal, closed in the 1970s. This was the last major traffic on the Canal.

Boats belonging to the Willow Wren carrying company round the turn at Hawkesbury Junction, Coventry, and meet an empty pair coming through the stop lock into the Oxford Canal.

findings were published in 1955 (the canals' equivalent to the 'Beeching' report).

Of the 2,101 miles of nationalised canals, only 336 miles – mostly of broad canals and rivers carrying coal, oil and steel products – were considered to be busy and suitable for

improvement. 994 miles could be kept because they were currently useful but had an uncertain future; 771 miles seemed to have no commercial use and should not fall under the nationalised system, effectively putting them at risk of elimination. Canals in the second group would be moved to the third group as their usefulness ended. From 1956 major expenditure began on the busy canals and the fight began to save those in the other groups.

In November 1944 a book *Narrow Boat* was published, in which its author, Tom Rolt, told gently of his travels through the English canals in the autumn of 1939. People read the book and a group formed round its author and then developed to become the Inland Waterways Association (IWA), continuing to gain members and support. Nationalisation helped; it was easier to lobby Parliament once canals were nationalised than it ever would have been if they had remained independent.

The IWA was to split as personalities clashed and people's ideas about the future of the canals varied: Tom Rolt was a believer in a way of life – canals as a business; Robert Aikman was for keeping them accessible at any price and opening them up as canals for all; Charles Hadfield was a pragmatist looking at what could be practically achieved. He would later

Narrowboats continued to carry gravel at Thurmaston on the River Soar in the 1980s. Many independent carriers and enthusiasts worked hard to keep the canals alive.

Coal loading at Plank Lane on the Leigh branch of the Leeds and Liverpool Canal in the 1960s. Carriers used ex-British Waterways craft.

join the British Waterways Board and work on the inside, and he also became the foremost historian of the waterways. Tom Rolt, having inspired a movement, went on to write more books and restore a railway, while Robert Aikman led the fight to save the group of canals given little future by the Rusholme report.

Nationalisation did not help the waterways improve their position in the transport hierarchy. With central control over the movement of goods, canals found themselves frozen out of some traditional cargo flows, which were now moved to road and rail.

By 1960 waterways transport was seriously affected by roads. Railways were also affected: the 1963 Beeching report on the railways, while closing many lines, recommended the introduction of bulk coal trains to larger power stations. Eventually many of the smaller barge-served power stations closed.

Oil began to replace coal as the fuel for factories, and this would not be delivered by water except on the broadest waterways. British Waterways' transport arm, while trying to maintain its role in waterway freight carrying, used its warehouses for more and more lorry traffic.

In 1962 the British Transport Commission was wound up. From 1963 canals would have their own body, the British Waterways Board. Reviewing their options, this group almost immediately decided that carrying on the narrow canals was unsupportable. By autumn 1963, virtually all the contracts for the northern and southern narrowboat fleets were terminated. Carrying on the Leeds to Liverpool Canal also mainly ceased, with only a few contracts continuing into the 1970s.

Some of the remaining boating families joined the independent Willow Wren, a firm of carriers established in 1952. The final bulk cargoes on the narrow canals would cease in 1973.

On the broad canals and rivers barge traffic continued. One final attempt to revitalise carrying in the north-east saw the Sheffield and South Yorkshire improved for steel traffic in the 1980s. A scheme long delayed by government finally opened as the bulk steel industry collapsed.

One last major canal traffic was the export of coal from Goole using the system of compartment boats developed in the Victorian era. In 1986 this traffic flow ceased, the equipment in use was at the end of its life and replacement with new machinery was not viable with the small volume of

A hire boat coming through detergent effluent at Bratch Locks on the Staffordshire and Worcestershire Canal, the start of a new age for the waterways.

A site visit to a repair stoppage on today's canal – a chance to see behind the scenes and attract new members and volunteers to the Canal & River Trust.

coal being moved. The last bulk movement of coal on water ended in 2003 when Ferrybridge power station ceased to take coal from the Aire and Calder Navigation.

After the main traffic flows on broad and narrow canals dried up, individuals carried on finding contracts and carrying cargoes, but even they have now mostly finished apart from bagged coal to canal-side customers and some sand and stone traffic.

On the narrow canals the way forward was seen to be leisure and community use. In 1968 a new Transport Act gave the remaining waterways security from closure and allowed for partnerships with local authorities to help develop a new life for the waterways.

As industry left the canal banks, water quality improved and canals were seen as green arteries into cities. The pioneering work of the British Waterways architects department under Peter White from 1970 showed how canals could become part of the modern city and how their heritage could be preserved while changing their use across the country.

Once traffic ceased, income came from water sales to industry for cooling; rents and property income; and private boats and hire boat companies. A licensing system for boats replaced most of the old toll structure, and government and local grants made up the shortfall.

The first hire-boat base had been opened on the canals at Christleton in Cheshire in 1935. After the Second World War as the canal system became public a hire-boat industry gradually opened up. New holiday boats were designed and built, more people came onto the water and more people became supporters of waterways.

In the 1960s and '70s, volunteers and the Inland Waterways Association showed a way forward; their free labour cleared canals that had filled with mud and rebuilt locks. The nationalised industry resisted at first but then eventually followed their lead and some closed stretches of waterways reopened. The National Trust and independent canal restoration societies took on further projects. In the 1980s impetus came from the Manpower Services Commission to give jobs to unemployed people, and from

Canals in the twenty-first century: the Falkirk Wheel on Scotland's nationalised canals, reopening the link between Edinburgh and Glasgow and an attraction in its own right.

the 1990s National Lottery funding has been applied to canal restoration.

Money generated from development and the lottery led to renewed dredging and maintenance through the 1990s and into the twenty-first century. More canals have reopened and major structures have been renewed.

In the wider world economic and political changes were happening fast. The global recession, which began in 2007, heralded a new era of uncertainty in the public sector and of universally tightening finance. By late 2008, the board of British Waterways was asking itself questions about how to ensure the long-term sustainability of the canals and rivers in its care. Government funding was reduced and waterways seemed unlikely to be high up a wish list that also contained hospitals, schools and roads.

A whole range of options was evaluated, including privatisation, but the idea of becoming a charity emerged as the favourite. Starting with a cross-party endorsed launch in the House of Commons in May 2009, British Waterways began to make the case for its own demise and replacement by a new waterways charity. In its discussion document *2020 – A Vision for the Future of our Canals and Rivers*, Chairman Tony Hales wrote, 'The private sector built the canals, the public sector rescued them and I believe the third sector can be their future.'

The challenge was to create a new and enduring framework that would allow the waterways to adapt successfully and sustainably over a long future. The Scottish government decided to keep their canals in public ownership. In England and Wales the specification had to allow for public funding when required and at the same time to allow the potential of supporters, volunteers and donors to be given free rein.

The Canal & River Trust was finally launched on 12 July 2012 to look after 2,000 miles of waterways in England and Wales. A public trust for the waterways came into being 250 years after it was first discussed. The Canal & River Trust will face huge challenges in the future but in many ways the waterways system that after the First World War virtually disappeared from view has climbed back into everyday life and has successfully achieved a remarkable rehabilitation.

FURTHER READING

Hadfield, Charles, and Joseph Boughey. *Hadfield's British Canals.* Sutton Publishing, 1994.

Hadfield, Charles. Regional histories. David & Charles, various dates.

McIvor, Liz. *Canals: The Making of a Nation.* BBC Books, 2015.

Paget-Tomlinson, Edward. *Britain's Canal and River Craft.* Landmark Publications, 2005.

Paget-Tomlinson, Edward. *The Illustrated History of Canal and River Navigations.* Landmark Publications, 2006.

Rolt, L.T.C. *Narrow Boat.* The History Press, 2014.

WEBSITE:

Canal Junction, www.canaljunction.com

THE PACK-HORSE CONVOY.   By Louis Huard.]

PLACES TO VISIT

Canal and River Trust (various sites nationwide).
Website: www.canalrivertrust.org.uk

Anderton Boat Lift, Lift Lane, Anderton, Northwich,
Cheshire CW9 6FW. Telephone: 01606 786777.
Website: www.canalrivertrust.org.uk/enjoy-the-
waterways/museums-and-attractions/anderton-boat-lift

The Canal Museum, 3 Bridge Road, Stoke Bruerne,
Towcester NN12 7SE. Telephone: 01604 862229.
Website: www.stokebruernecanalmuseum.org.uk

Falkirk Wheel, Lime Road, Tamfourhill, Falkirk FK1 4RS.
Telephone: 0870 050 0208.
Website: www.scottishcanals.co.uk/falkirk-wheel

Fourteen Locks Canal Centre, Cwm Lane, Rogerstone,
Newport NP10 9GN. Telephone: 01633 892167.
Website: www.fourteenlocks.mbact.org.uk

Linlithgow Canal Centre, Manse Road Basin, Linlithgow,
West Lothian EH49 6AJ. Telephone: 01506 671215.
Website: www.lucs.org.uk

London Canal Museum, 12–13 New Wharf Road, London
N1 9RT. Telephone: 020 7713 0836.
Website: www.canalmuseum.org.uk

The Museum of the Broads, The Staithe, Stalham NR12
9DA. Telephone: 01692 581681.
Website: www.museumofthe broads.org.uk

National Waterways Museum, South Pier Road, Ellesmere
Port, Cheshire CH65 4FW. Telephone: 0151 355
5017. Website: www.canalrivertrust.org.uk/enjoy-
the-waterways/museums-and-attractions/national-
waterways-museum

National Waterways Museum Gloucester, Llanthony
Warehouse, The Docks, Gloucester GL1 2EH.
Telephone: 01452 318200. Website: www.canalrivertrust.
org.uk/enjoy-the-waterways/museums-and-attractions/
national-waterways-museum-gloucester

Standedge Tunnel and Visitor Centre, Waters Road,
Marsden, Huddersfield HD7 6NQ. Telephone: 01484
844298. Website: www.canalrivertrust.org.uk/enjoy-the-
waterways/museums-and-attractions/standedge-tunnel-
and-visitor-centre-west-yorkshire

Waterways Ireland Visitor Centre, Grand Canal Quay,
   Dublin, Ireland. Telephone: +353 1 677 7510.
   Website: www.waterwaysirelandvisitorcentre.org
The Yorkshire Waterways Museum, Dutch River Side, Goole
   DN14   5TB. Telephone: 01405 768730.
   Website: www.waterwaysmuseum.org.uk

INDEX

Page numbers in *italics* refer to
   illustrations

Aberdeenshire Canal 54
Aire and Calder Navigation 5, 6,
   8, 18, 26, 47, *47*, 54, *55*, 56, 57,
   63, 73, 74
Andover Canal 54
Avon (Warwickshire) *15*, 18
Birmingham Canal Navigations 31,
   32, *40*, 41, 52, *52, 64*
Bridgewater Canal 20, 21, *22*, 24,
   41, 53
Brindley, James 21, 22, 23, 24
Calder and Hebble Navigation 26,
   26, 27
Caledonian Canal 34
Cam, river 11
Carr Dyke 11
Chesterfield Canal 47
Coventry Canal *19*, 24, 26, *41*,
   70
Crinan Canal *34*
Douglas, river 18
Duke of Bridgewater *20*, 21
Edinburgh and Glasgow Union
   Canal 34, *75*
Ellesmere Port *4*
Exeter Canal 14
Forth and Clyde Canal 33, 63, *75*
Foss Dyke 11

Glasgow Paisley and Johnstone Canal
   44, 45, 46
Gloucester *7*, 12, *14*
Grand Union Canal/Grand Junction
   Canal 6 ,*7*, *8*, 29, *30*, *35*, *37*, *38*,
   40, *43*, *51*, 53, 54, *58*, *59*, *61*, *62*,
   63, *65*, 66, *66*, *67*, 69
Great Ouse, river *9*, 12
Horses 36, 37 38, 39, 40, 58
Huddersfield Narrow Canal *42*
Icebreaking 37, 38, 39, *40*
Ireland 35
Jessop, William 29, *31*, 34
Kennet, river 16
Kennet and Avon Canal 28, 29,
   53, 63
Lee, river 13, *36*
Leeds to Liverpool Canal 5, 6, 26,
   27, *27*, 69, *70*, *72*, 73
Macclesfield Canal *38*
Medway, river *18*
Mersey, river 18, 19, 21
Mersey and Irwell navigation 21
Monkland Canal 33
Monmouthshire and Brecon Canal
   33
Navvies 25, *26*
Ouse, river Yorkshire 5, 13, 18, 26
Oxford Canal 24, 26, 29, 53, 54, 70
Packhorse 19, *77*
Regent's Canal 35

Rennie, John 29, 49
Rochdale Canal 69
Sankey Canal 19
Scotland 33, 34, *44, 45*, 46, 54, *75*
Severn, river 7, 12, 14, 16, 17, 24,
   55, 56
Shropshire Union Canal/
   Birmingham and Liverpool
   Junction Canal 31, *37*, *39*, 41,
   42, *43*, *50*, 51, 52
Soar, river *71*
South Wales 31, 32, 33
Staffordshire and Worcestershire
   Canal 24, *25*, 29, 55, 69, *73*
Stoke Bruerne *6, 7*, 50
Stour, river (Suffolk) 16, 17, *17*
Stover Canal *47*
Telford, Thomas 31, 34, *34*
Tennant Canal *33*
Thames, river 12, *16, 17*
Thames and Severn Canal 26
Trent, river 7, 12, 61, 66
Trent and Mersey Canal 7, *9*, 23, *23*,
   24, 29, *32, 51, 68*
Weaver, river 7, 19, 24, 57, *57*
Wedgwood, Josiah 23, 24, *24*
Wey and Arun Junction Canal 61
Witham, river *13*
Worcester and Birmingham Canal
   56, *64*